THE SENSORY LANDSCAPE OF CITIES

BY CHARLES LANDRY

02

First published by Comedia in the UK in 2012
Copyright © Charles Landry

ISBN: 978-1-908777-01-0

Comedia
The Round, Bournes Green
Near Stroud, GL6 7NL, UK

Book design: **www.hillsdesign.co.uk**
All photographs: **Charles Landry**
Cover photograph: *Tobias Rehberger's café which won
the Golden Lion at the 2009 Biennale, Venice.*

This and other Comedia publications are available from:
www.charleslandry.com

*Barcelona: The famous
La Boqueria food market
on the Ramblas.*

CONTENTS

Ray-Ban
GENUINE SINCE 1937
GINSA
PUBLICIDA EN FACHADAS
900 50 13 18

SUMMARY

This is the second in a series of short publications, which seek to briefly encapsulate key agendas and thought movements that shape cities today and will have an impact on their future.

The city is a communications device. It speaks to us through every fibre of its being. The lived urban experience comes from a circular sensory cycle. *The Sensory Landscape of Cities*, sees the city as a 360-degree, enveloping, immersive experience, which has emotional and psychological impacts. It argues that we sense, feel and understand it through increasingly narrow funnels of perception. Living in an impoverished perceptual mindscape makes us operate with a shallow register of experience and understanding about what is important for our cities to survive well. At the same time, commerce and media seek to pump up our desires in ever more shrill tones. We have more sensory bombardment, but have lost the art of appreciating the varied sounds, smells, the texture and quality of materials and the look and feel of the city and its component parts. There is distraction, loss of attention, concentration and focus. Sensory appreciation builds the knowledge upon which our worlds are built. This effects beliefs, choices and our priorities for change.

There is a desire in many to take more control of our sensory environment, rather than being passive consumers or even victims of its current effects. Whether we allow commerce and the media to trespass on our senses will be one of the crucial battles in defining how our cities evolve in the future. Yet the urban policy world has insufficiently understood the power and potential of the senses. The contrast between everyday sensing of the city and how policy-makers usually describe cities as lifeless beings is stark. A greater understanding of the importance of environmental psychology is crucial. This focuses on the interplay between people and their surroundings and the degree to which it creates stress or feels restorative.

The city is a 360-degree, enveloping, immersive experience. It has emotional and psychological impacts.

bao: Vast
lboards
creasingly invade
blic space. Is this
autiful or ugly?

PREFACE

This essay summarizes and takes forward elements I have explored before. In *The Creative City: A Toolkit for Urban Innovators* (2000), I began to describe how planning should change and become a much more holistic endeavour, combining the insights and expertise of those concerned with the physical hardware of the city, like spatial planners, architects or urban designers, as well as those concerned with the soft and intangible matters.

These might range from anthropologists to social activists, to economists, ecologists or people working with culture.

I suggested that a vital part of city making should be for decision makers to undertake a 'survey of the senses' so that they understood much better the lived experience and emotional life of the city and its impacts on peoples' well-being and psychological states of being. Physical or spatial planners often see the city from the air rather than from eye level and the ground. Architects often merely think of covering three-dimensional space. Together they often dominate discussions on the city.

I proposed too that any planning activity by individual project developers or the city itself should take a 360-degree perspective by looking at the city 'through the eyes of ...'. My attempt was to shift focus away from seeing planning as a largely technical discipline dominated often by middle-aged guys; interesting and competent as they might be. Young or old people, men or women, rich and poor, business interests or social activists, existing inhabitants or newcomers and immigrants experience the city in varied ways, yet we often pretend there is just one view. The joy and difficult art of city making is to immerse oneself into this complexity and to try to find solutions that at least find some balance between potentially conflicting aims and world views. In *The Art of City Making* (2006) I take a number of sensory journeys through places across the globe using each of my individual senses. Travelling through a city with an alert nose reminds us of the differences. The salty or sour sea smell of a harbour city might

London: A morning
rush at Paddington
station with too
much information.

Dubai: A metro ride through an urban jungle.

also give you a fresh high from its ozone. Contrast this with the hot and moist smell of a tropical city. A polluted city after a while gives you a sickly, sweet headache. Your emotions feel different in each. The same is true if you travel the city awake to its noises, or what it looks like or what its texture is. Every city has its own sensescape.

Every city has its own sensescape.

Many literary figures and cultural critics have described the city in this richer way, but their insights have not become part of mainstream city making.[1] So far, a way has not been found

[1] Klaske Havik's as yet unpublished doctoral dissertation *Urban Literacy: A Scriptive Approach to the Experience, Use and Imagination of Place* (2011) is a good summary of the increasing literature see: nl.linkedin.com/pub/klaske-havik.

organizationally to incorporate their knowledge on a consistent basis and if we do it is usually as an 'interesting' add-on. There remains a sharp division between those with insights from the humanities and those with a background in science, technology and engineering. To have a more complete view of cities and how to turn spaces into places that develop and exude meaning, identity and pride, requires a realignment, mutual respect and understanding of what each party can bring. The danger is that we may build urban infrastructures, such as roads, that work functionally, but feel unsatisfactory or over-engineered for those that live near them and have to negotiate them in their daily lives.

SETTING THE SCENE

The city is a 360-degree, all-embracing, immersive experience, but we sense, feel and understand it through increasingly narrow funnels of perception. What might that be? Is that at 90 or even only 45 degrees? Our personal sensory landscape is shrinking, precisely at the moment when it should broaden, whilst the world of commerce in the name of experience, attempts to grab our sensory attention and pump up our desires, in increasingly strident ways.

Our perceptive capacities are cramped because we do not sufficiently recognise or practice most of the senses. We are losing our ability to distinguish noise registers, rhythms and pitch, masked by the continuous low-level din of traffic. We do not give ourselves time to appreciate the subtleties and qualities of materials, such as whether they are soft, warm and yielding, or harsh, cold and unforgiving. Or even how the patina of the ages seeps out and lets a building silently speak. We crave the rich smells of the complexity of freshness as you enter a market at the fruit and vegetable stalls; as these feel 'authentic'. More frequently instead, we encounter the powerful, heady blast of perfumes and cosmetics as they seep through department stores; or their waft of warm stale air in colder climes, and in warmer ones, a draught of cold.

We have forgotten our 'natural' aesthetic knowledge, which is the science of sensory perception deriving from the Greek word meaning to be 'sensitive' and 'sentient' and to 'perceive', 'feel' and 'sense'. Aesthetics is about being able to discriminate the good, the bad and the indifferent at a sensual level. We need to stretch our aesthetic awareness so we can discern the important things that matter and deal with them. In Mediterranean cultures this understanding is often stronger than in Anglo-Saxon ones. People in the former touch foods and feel materials more readily and their language, such as Italian, has a more expressive vocabulary to describe the sensory realm.

nisoara: The
vers of history
ep through the
ilding.

The sensory landscape of a city is the totality of experiences you perceive by being in and navigating a city. Not only the obvious ones such as, seeing, smelling, hearing or imagining the feel of the textures you encounter. These are the tangibles. The landscape includes invisibles such as electroperception. The city is a vast, dense sea of electrical energy fields and waves, estimated to be 100 million times stronger than they were 100 years ago. These massive currents criss-crossing the urban environment are unseen, unfelt, unheard; without taste or smell. The accumulative cocktail of magnetic and electrical fields generated by power transmission lines, pylons and masts, mobile phones, computers, television and radio, lighting, wiring and household appliances, can seriously interfere with the subtle natural balances of each cell in our body. The sum of visible and invisible perceptions is experienced emotionally. How we experience these sensations determines the mood, disposition, temperament and ultimately personality of a city. In an iterative process the effect of these tangibles and imperceptibles shapes and reinforces how we perceive the city, how we behave in a city and what it becomes.

By looking, sensing and feeling closely we can discern the overall atmosphere and specific signals of a place. The city then embraces us fully. These sensations are often encapsulated in a feeling of 'yes' or equally a response of 'no'. Blank walls, the inability to walk, relentless asphalt, or glass reflecting back at you, project a 'no' reaction like 'keep out' or 'stay away'. A barren, sterile building made up of unforgiving, cold materials can trigger reactions that the world is bleak and wanting. An acrid smell can make you tighten up and shrivel into yourself. A penetrating noise can stop the pleasures of our meandering thoughts.

A good secondary shopping street, by contrast, with its diversity of often very local shops, interspersed perhaps by a pocket park, tends to give you a sense of 'yes' since your eyes are diverted and entertained. The street says you can join in. By observing with all senses, calmly alert, you can figure out whether people you encounter seem to be close, respectful and open or suspicious, tense and rushed. Without even speaking to anyone we can sometimes measure the pulse of conviviality – how people get on. Equally we can take in urban decline or growth, whether a place is loved or cared for, how polluting it is, how powerful it is.

By looking, sensing and feeling we discern the overall atmosphere and signals of a place.

Chicago: An unusual view of Anish Kapoor's playful sculpture in Millennium Park.

With grids and blocks you get a better sense of the near and the far and with curves, bends and loops a better sense of the more hidden and intimate. Depending on the city we want, we privilege one over the other. This is what we might call urban literacy – the understanding of how places work.

Crucially, we need language to make this experience conscious and real, as a precondition to act upon the consequences. Yet, the language of the senses is not rich enough for describing our cities today. Our language, unless we look to literary figures or artists, is hollowed out, eviscerated and dry. It is shaped by the technical jargon of the professions, especially those of planning and the built environment. The language to describe the urban look is dominated by the physical, without descriptions of movement, rhythm, or smell and sound. This visual language comes largely from the architecture and urban design professions who still dominate discussion on cities. Descriptions of the visual city come from habits of portraying classic architecture where building

Our language for the senses is not rich enough to describe our cities.

13

components are illustrated: Pedestals, columns, capitals, pediments and architraves. The language has broadened somewhat, yet still with a focus on static elements rather than dynamic wholes. Urban design, meanwhile, sees and describes cities more as dynamic totalities: Place, connections, movement, mixed uses, blocks, neighbourhoods, zones, densities, centres, peripheries, landscapes, vistas, focal points, and realms. But, both frequently exclude the atmospherics of cities, the 'feeling' of the look. Does it make you shrink into yourself, make you calmly reflect or fill you with passion? Does it close you in or open you out? The noise is essentially that of mechanical devices and mostly cars. Stand for a moment in a city and try to hear the 'sounds of silence'. Silence is a rare commodity.

Bilbao: This inappropriate health services building scrambles the brain.

Smell, rarely discussed, is extremely evocative, as evidenced by neuroscience. Today, smells are dominated by petrochemicals in the urban realm; whose polluting effects have dramatic health impacts. In the past it was the nauseating and putrid smell of faeces and waste from animals and humans that dominated, and the overwhelming noise of horses clacking on cobbles. Yet smells can relax and heal. Recently in Aberdeen, I arrived overloaded and uptight to give a talk in the Botanical Gardens. To get to my destination I had to walk through a 30-metre-long scented garden. I came in tense and left the space floating. These are the intoxicating effects of good smells.

Did we ask for this soundscape, smellscape or visual appearance?

Sensory awareness reminds us that we can ask: 'Did I ask for this soundscape or smellscape or this visual appearance?' 'Was I consulted or does it just happen?' Trespassing on our senses will be one of the critical battles in defining how our cities evolve in the future.

OVERLOAD & DISTRACTION

Increasingly our primary sensation in cities is one of overload. It makes you feel that things are out of control. Cities feel as if they are running away with themselves, where the constant stimuli of images, sounds, words, take on a life of their own. Uncontrolled apparently, but with an inner logic, we experience the result as white noise.

The emotional impacts and psychological effects of being in this kind of city are strong and largely negative. They distract, fragment and can shatter our attention. Take advertisements. There are ads on airline sickness bags, on bannisters, railings, stairs, escalators, in toilets, on tables, on chairs, on floors, on sidewalks and on complete buildings. The urban landscape at times appears like a mosaic of adverts. What is more, they are being increasingly converted into digital screens heightening the sense of motion, fizz and clutter. You cannot escape an ad and they are getting bigger. The Lagoons ad in Dubai was 400 metres long and 20 metres high and the Betfair.com ad for the Euro 2008 football championships measured 50 football fields in size. The feeling of ubiquity or all pervasiveness is fuelled by our experiences elsewhere, from always online smart phones and ever-present spam e-mails or brand names featured in TV shows and films. The market research firm Yankelovich, estimated in 2007 that a person living in a city 30 years ago saw up to 2,000 advertising messages a day, compared with up to 5,000 today. The paradox is that the freedom of choice projected through ads as liberation, can then be experienced as claustrophobic and crowd you in. It is largely a question of scale and extent. Of course some better adverts or hoardings are like art installations and more attractive than the spaces they cover up or disguise. Adding to the overload are the counter messages provided by ever-present graffiti. At times amusing, clever and delightful, it is often merely an angry or ugly defacement; where urban tribes mark their territory.

What will the effect be on the new generation, who have never experienced anything different and are unaware of sensory richness?

ver: *Invasive*
ertising in an
s complex.

rleaf:
resting graffiti
mentary in
ge and buses as
erts in Tirana.

Italia në shtëpinë

UNIVERSReklama
INDOOR & OUTDOOR ADVERTISING

SIGMA
VIENNA INSURANCE GROUP

CHRONIC MYOPIA

We approach the urban sensescape with chronic myopia and thus an ill-equipped toolkit. Paradoxically, this aggravates the problems, dysfunction, or malaise we might wish to solve. This feeling of not sensing can dull and foster an atmosphere of being out of control, taking people almost to breaking point.

A focus on senses is not to make people feel hyper-aware or over self-conscious. Instead, it aims to get us to concentrate on two things. First, how we feel as individuals and city-dwellers in negotiating urban life, so as to live well, generate wealth, coexist without harming fellow citizens and to collaborate with them. Second, to care for the environment without which our life is not possible. This expanded awareness is far-reaching. It demands, unavoidably, that we collectively take more control over, or guide our sensory environment, rather than let commercial forces mediate, shape and determine our perceptions of the world. It is now a question of balance and assessing the right level. Clearly, Eastern Europe in the Communist era was visually far too dull and the public ads brought life. Now there is a groundswell of opinion emerging that questions the value of overweening commercial messages, like vast billboards. It is bitterly opposed by the outdoor advertising industry. As I write this the first study on the economic impacts of billboards on nearby real estate value has come out: *Beyond Aesthetics: How Billboards Affect Economic Prosperity* by Jonathan Snyder. It shows billboards negatively affect the values of neighbouring properties, and cities with stricter billboard controls experience greater economic prosperity. Thus many hoardings covering heritage sites under restoration, state they are contributing financially to the renewal.

This battleground is extending to intrusions outside of the public realm, such as hidden adverts in films or unwanted sales calls. In the future people will pay not have interruptions. For example, the music website Spotify - service is free with adverts and costs a fee without. Peace will be a luxury item.

The counter messages, such as how our behaviour and lifestyles might need to change in order to address complex urban problems,

ala Lumpur: e oversized velopments can owd in on us.

Seoul: A public demonstration reminds us how many people live in megacities.

Seoul: Traffic before dawn in downtown.

struggle in this environment. Creating the necessary shift is a political question.

Another dimension is to try to experience and see the world with less sensory impositions by commerce, the media, by those who make cities or even our own cultural habits. This aim is to trigger a more direct, unmediated view and response (while noting that nothing is completely unmediated). The challenge is to see without pre-judging and to perceive before reflecting, and getting rid of the sensory clutter makes this easier. The process of unravelling and folding back is never complete. Yet it is a worthy endeavour as it helps us be more in control and more authentic to ourselves. The idea of authenticity is contested. We are always shaped by cultural norms and experiences and Sartre's, 'absolute freedom' is fanciful. Erich Fromm notes, there is enlightened and informed motivation that acknowledges influences, critically accepting or rejecting them as appropriate. This form of authenticity gives the chance to ease into sensory perceptions with more openness. My experience is that this approach helps clarity.

... as cities grow they can become more claustrophobic.

INVIGORATING SENSES

Seeing the city as a field of senses can be invigorating. Playing with them can trigger action; it might generate the pressure for ecological transport more quickly, for planting more greenery, for creating spaces where different cultures want to mix or for balancing places for stimulation and reflection in the city.

It would force us to ask questions such as: How can the smell, sound, visual, touch and taste landscapes, help cities? Bold inroads into sensory fields have already been undertaken by some cities. Light and colour have been tackled. Lyons in France and Ghent in Belgium are renowned for their lighting plans and light festivals and they get three Michelin stars for their efforts. Colour planning strategies exist in places such as Norwich in Britain or Italian historic towns. Rarely though are effects on the mind and wellbeing considered. This will increasingly become a planning issue as colour and digital screens on buildings are increasingly becoming part of the repertoire. Imagine the differences in effects of a city that is essentially white (Casablanca or Tel Aviv), pink (Marrakesh), blue (Jodphur or Oman's new Blue City project), red (Bologna), yellow (Izamal in Yucatan). Or imagine a city that is black – the darkness would provoke Seasonal Affective Disorder (SAD), well-known in the Nordic countries where winter light is scarce. Until the 1960's, London was in fact a black city. Emissions of smoke from coal and industry, blackened stone and brick, shading buildings with a uniform, light-absorbing black. Decades of scraping off the surface dirt reveal colour and detail hidden for years. This was like a mental liberation. The nickname of some cities involves colour: Berlin or Milan are both known as 'the grey city'.

Clearly planners and developers deal with sensory elements, but not from a holistic perspective and thus with insufficient insight. The urban design view does not usually consider environmental psychology. There are, thus, strict standards and codes for signage, height configurations or use mixes, yet the effects of dull looking roads are rarely considered or challenged. Perhaps this is

rth: A stylish
w development
the city centre.

*Düsseldorf:
Frank Gehry's
building
triggered
the harbour
regeneration.*

Our challenge
is to create
physical
environments
that feed the
spirits.

because of the power of transport departments whose sensory sensitivities are not renowned. Yet they have had an impact on advertising hoardings. How many will they allow, where can they be or how large? Sao Paolo, from being a city with vast hoardings lining their canyons of skyscrapers, has now banned them all. Kolkata in 2011 reduced billboards by 50 per cent. Moscow, after a period of allowing adverts to invade every crevice of the city is now considering much tighter controls. How do you read the city without commercial signs? Italy was the honking capital of the world until it was banned 20 years ago. In late 2011, New York taxis were warned about blowing their horn: 'Drivers remember that honking is against the law and liable to a $350 fine except when warning of imminent danger'. Each country has a different response to the blowing of car horns. In Britain it is not part of the culture, but by contrast in Turkey horn honking is continuous, but not personal. It happens even when a vehicle is 20 rows back from the green light. Even motorbikes give an impotent little honk, very much a nasal 'eeee, eee', like a sheep with sinus problems.

Our sensory awareness is perhaps most strongly manipulated by the world of ubiquitous shopping malls and their destination marketing whose purpose is for people to spend more. These worlds project two primary images: The vast car parks outside and the bright shiny world inside that pulses, flickers and shouts its messages at various levels of shrillness. Enticing imagery or 'nice' smells and 'good' sounds direct and guide people. Yet at a minimum we should know what is happening. For instance, the artificial smell of bread is pumped out in supermarkets, as is the smell of turkey at Christmas. The perfume smells seeping out of stores is manufactured to create moods receptive to buying.

The challenge is for messages with a higher purpose, such as concerning climate issues or wealth distribution, to get through given the sophistication, imagination, professionalism and resources mainstream industry can work with. Freiburg's sign system is a good example, where the name of settlements reveal their ecological intent without hectoring, such as with the 'Sonnenschiff' (Sunship) settlement or hoardings that proclaim 'Hinter dieser Tür arbeiten kluge Köpfe' (Behind these doors clever people are working) so reinforcing the city's knowledge city intentions.

Paris: Curvy hoarding disguising a building off the Champs Elysees.

MINDSET & MINDFLOW

Creativity has a general, all-purpose problem-solving and opportunity-creating capacity. Its essence is a multifaceted resourcefulness and the ability to assess and find one's way to solutions for intractable, unexpected, unusual problems or circumstances.

There is logic, however, to looking and perceiving things with a restricted mindset. Blocking things out is a rational thing to do. If we were to be completely open all the time we would go crazy. It would be disorienting. There would be too much information – no place to anchor ourselves. We would be swimming within endless impressions that could overwhelm us – their full blast could mean we are experiencing too much. Yet, there are times when we want to open out to fuller awareness and greater appreciation of things. This time is now.

Mindset is the order within which we structure our worlds. It is the settled summary of our prejudices, those matters where we have already pre-judged, and the priorities and rationalizations we give them. This reflects our choices, both practical and idealistic, based ultimately on our values, culture, philosophy, traditions and aspirations. Mindset is our convenient, accustomed way of perceiving, looking, filtering and thinking about things. It not only governs what we notice in wandering the city and interacting with it, it also moulds what we see or think about and interpret in places. Mindflow is the mind in operation. The mind is locked into certain patterns for good reason. It uses familiar thought processes, concepts, connections and interpretations as a means of filtering and coping with the world. The brain hears what it expects or even wants to hear; sees what it expects to see and discards what does not fit in. This operates below the level of conscious awareness.

Mindset and mindflow create the mindscape. This is the totality of our thinking: the modes, proclivities and gut reactions of thought; scattered intuitions, emotional responses, the theories we use to interpret and construct reality. In turn this shapes all the sensory elements and how these are perceived, taken apart

*Kyoto: Ginkaku-ji
or Silver temple
Zen garden.*

and interpreted. How our mind responds comes from within us and is also moulded by the media and cultural representations. This mind sets the preconditions for our perceptual geography, which is the process of acquiring, interpreting, selecting, and organising sensory information about the places we inhabit.

The sensory realm of cities is not neutral and value-free. It is subjectively interpreted, yet similar emotions are often shared, especially between individuals within a cohesive group. Conversely, whilst having emotions is universal, our culture determines how our emotions unfold and how we interpret their significance, as do expectations, norms and the conditioned behaviour of the group.

People within and between cultures perceive and value the senses in different ways. This means we may be walking through a place and having completely different experiences. A Chinese, Italian or Arab will have varied expectations. We have differing thresholds of acceptance. Some cultures seem to prefer and enjoy noise and hubbub. Contrast a market scene in the Middle East or Africa with one in the Nordic countries where silence is more valued.

Noise may feel less loud or the colours less brash for the Chinese. In another culture they may be screaming. The sensory environment for an older person might feel noisy

or unsafe, and too quiet or safe for someone young. The same differences can apply to people from different classes and income backgrounds. A smell is seen as sweet and comforting in one cultural context and as fear inducing in another. A smell can be nice if you associate it with someone you like, horrible if exuded by someone you dislike. The sound of nothingness may feel relaxing to a Finn and like a heavy rumble to someone from Taipei. And for each of the landscapes of sense, there are cultural codes of conduct. The Chinese and Italian speak far more freely about smell in comparison to the English. Southern Europeans are encouraged to touch merchandise, especially fruit or vegetables, whereas it is discouraged elsewhere. In Northern Europe, people tend to touch each other less. Southern Europeans shake hands, hold shoulders more and embrace.

Ask yourself when you last really noticed a city simultaneously - both emotionally and intellectually. What colours did you see? What shapes? Did they look different depending on the light? What was the effect on your mood of different colours? How far could you see? Was the scene opening out or closing in? What did you smell? Can you recall them? What sounds do you hear? Did you hear birdsong or a dog bark? What did things look like, such as the pavement you walked on or the road you crossed? What materials were they made of? What did you touch? Were the materials harsh? Recalling like this is usually easier when travelling - the unfamiliar often sharpens the senses.

BLOCKING OUT EXPERIENCE

Why is there so much blocking out? Sensory manipulation and overload is perhaps the primary experience of the 21st century large city dweller. Look closely at people and especially try to see their eyes. You sense increasingly that of necessity we shield ourselves from experiencing. We put the shutters down and so the commercial world screams more loudly at us.

We walk city streets with reduced peripheral vision. We look straight ahead or down and on occasion up. Our eyes dart sideways left or right. We tend to rush. Rarely do we take in the rounded view - slowly. In part this is because we are walking faster. A study in the 1990's and compared again in 2007, showed that on average we are walking 10% faster than we used to, with Singapore (30%) and Guangzhou (20%) having the greatest increases. But Copenhageners and Madrilenos walk pretty fast too. The most sluggish walkers are in the Middle East. Walking speeds have many reasons: Perhaps frenetic expectations in Singapore, the cold in Denmark or being late in Madrid and the heat in the Middle East.[2] Indeed, we are also eating, talking and communicating faster than we did 30 years ago, in an attempt to fit more into a 24-hour day. Squashing more into a tighter time frame makes quantity a substitute for quality, as the experience shallows out and is less remembered. More means less, and less can mean more.

Observing slowly is likely to happen in a park or sitting in a café. Speed apart, and the constant distractions of the ping of the mobile phone telling you there is another message, there are other reasons for closing in. Too many shops signs, adverts, public messages, instructions, exhortations, traffic notices – an information clutter, a spaghetti of impressions folding into each other. The messages are contradictory and difficult to sort. Sometimes the physical fabric of a city centre is simply a backdrop that asks us to buy, in ever more sophisticated ways. Simultaneously there are instructions telling you to stop at traffic lights or to walk, but many things need not be said. The physical

sbane: View
o the city
oss retrofitted
ial housing.

[2]http://www.richardwiseman.com/quirkology/pace_home.html

Beijing: A night market. Visually stimulating?

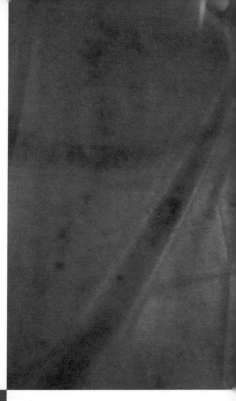

To avoid the assault on the senses we distance ourselves from the places we inhabit.

Things are often happening so fast everything becomes a blur.

fabric of the city such as barriers or closed entrances, tell you what you can and cannot do. At its best and on rare occasions, the sum total can be invigorating and feel like an art installation.

The city has many parts and the sensory overload is more evident at its core, the city centre, where functions and especially retailing coalesce. Yet every part of the city like suburban housing settlements, neighbourhoods or industrial estates or road systems, each have their sensory landscape and emotional effect. In one they seem vital and vibrant and in another deflating, dull and bland. The blandness of endless asphalt and concrete and cars is especially dispiriting.

In trying to avoid the assault on the senses we tend to distance ourselves from the places we inhabit. We enclose ourselves in our own worlds often literally, with our headphones on or go home and create our own private world. In so doing we are losing our visceral knowledge of cities. We can become less of a citizen by forgetting how to understand urban smells, to listen to its noises, to grasp its visual messages or be aware of its materials, exacerbated by multi-tasking, say walking and talking on the mobile phone. We are 'here' and 'there' at the same time which dissects attention. This may have new advantages of connectivity and networking, but better not to lose our primordial sensory knowledge in the process.

LIFELESS POLICY LANGUAGE

Cities are sensory, emotional experiences, for good and for bad. But we are not used to talking about them in this way: The smelling, hearing, seeing, touching and even tasting of the city are left to travel literature, brochures or novels set in a certain city.

Those that consider themselves to be the shapers and makers of cities, such as planners, surveyors, engineers, architects, economists or those that deal with its problems like social workers or housing officials, or those in the support services such as the police, talk of the city in narrow technical terms. It is as if the city were a lifeless being. This 'objective' lexicon is deprived of sensory descriptors.

City plans use the deadening language of - strategy, plans, policies, inputs, frameworks, spatial development code, development strategy, outcome targets, site option appraisal process, stakeholder consultation, statutory review policy programme, framework delivery plan, sustainability-proofing, benchmarking, empowerment, triple bottom line, mainstreaming, worklessness, early wins, step change, additionality.

There are rarely adjectives in these texts that might describe the qualities of a city. Using adjectives helps write the poetry of the city, which is an important way of putting the senses into a reality that can communicate. For many years I have asked people to give me adjectives or describe what they want their city to be and their likes and dislikes. The words they use are interestingly contradictory. They sound seemingly paradoxical. Some are lively, some calm, some excitable, some reassuring. The great city, of course, nourishes varied and opposite moods and moments. This makes it rich, layered and abundant. It is not either this or that, it could be both in a given time period.

an Francisco: The
mains of the
verhead highway
amaged in the
989 earthquake.

URBAN DESIRE

You have arrived – then depending on your frame of mind or spirit, you might want to be curious, discover something, explore, or even feel elevated, as if there is a higher purpose. In a less religious age the deep yearning for the spiritual remains. The desire for surprise is common and that surprise could be humorous and playful. Overall the sense is dynamic. It could be experiments in public space, such as a weird street performer making you see the ordinary in an extraordinary way or Aalborg's 'red carpet' day, where an extremely long red carpet glides through the city, or in Times Square where they periodically close off Broadway and scatter deck chairs randomly around the street to see what happens next. It has been a resounding success. These activities might simply stimulate you visually or make you see and experience the city differently.

People enjoy arousal once it is cradled and anchored in the familiar so you feel a sense of safety and security, which gives you the courage to explore. The city's public realm then is a home from home or an open air living room. This engenders relaxation and receptiveness so that the unusual is approachable. This makes crossing barriers and boundaries easier, whether it is mixing with a person from a different cultural background or going into a previously unknown area or street. New sense awareness comes into play in approaching the novel; your brain gets into gear. Keen attentiveness is key. It is a different kind of alertness from looking at the already known. Just looking has a passive element; but feeling passively at ease opens people out and releases a fuller range of senses. Impressions can be caught or even snared. Your peripheral vision is activated and this wider register allows greater subliminal thought and understanding. This processing

uala Lumpur: The
etronas Towers –
n icon competing
r world attention.

39

Paris: The dome of the Galeries Lafayette.

and interpretation goes on quietly in the brain, but you respond to what you see. If you then walk with this attitude of mind, it reflects itself in your gait – the pace you walk. Your experience of your own senses projects itself onto the rest of the city. This aesthetic understanding can have massive downstream effects in shaping our politics, choices and priorities.

Some words can make you feel the city viscerally; they seep through into your every fibre. Summing up what people desire and want from their city experience or what they dislike, gutsy, expressive words and phrases are used such as lovable, joyful, vital, captivating, 'a can-do place' or ugly, nasty, hideous, horrid.

The best public realm can be a home from home or an open air living room.

There is a calmer vocabulary too – beautiful, human scale, 'spaces which are places', balanced and harmonious. They use phrases like – 'I want to fall in love with my city', 'I want to be proud of where I live', 'it should be memorable', which tends to be the big hope. On occasion people bemoan the end of darkness through light pollution so that you cannot see the stars anymore. There is a yearning to reconnect to nature. To feel complete urban dwellers also want its opposite. Whatever the words it shows the city has etched itself into the mindscape.

Yet have you ever read a report about cities from an official body that uses this vocabulary? The contrast between everyday sensing of the city and how that becomes abstracted into lifelessness is stark.

IMPOVERISHED AWARENESS

Taken together is it not surprising that our direct encounter with the city and how we describe things is experienced at a low-level of awareness. We do not recognise, let alone portray, the smellscape, soundscape, the visual spectacle, the tactile texture, and the taste world sufficiently.

We articulate things in an impoverished way and it is made all the worse because the city can overwhelm our senses – honking, flashing, whirring, whizzing, precipitous, huge, confusing. Unsurprisingly, urban stimuli induce a closing rather than opening out. We feel depleted, drained and defensive and so our field of experience diminishes.

There are two sides to this interactive experience of person and place. What the urban environment throws at us and what perceptual range and antenna we throw at it. How much we make of this depends on the perceptual register we are operating at. Is it shallow or deep, is it alert, has it got the capacity to interpret impressions, stimuli, insights or conversation?

Living in an impoverished perceptual mindscape makes us operate with a shallow register of experience. This guides our lives through narrow reality-tunnels. By diminishing our sensory landscape, we approach the world and its opportunities narrowly. Consequently we do not grasp the full range of urban resources or problems at hand, their potential or threat, let alone their subtleties. We do not connect the sensory to the physical and thus work out how each can support the other or how the physical might be reshaped.

Our world is shrinking as its interconnections become far more tightly bound in, as mass movement and mobility continue unabated, as economies intermesh globally, and as electronics flattens the distance between places. This is happening at speed and simultaneously, rapidly bringing together cultures, people and ideas. To handle this complexity we need deep and discriminating minds that grasp the delicate diversities and understandings required to operate in worlds of difference and distinctiveness.

ragoza:
igmatic
ulpture by
ume Plensa
Expo park.

Constricted, we understand and interpret the city through the technical rather than the sensory, yet it is the sensory from which we build feeling and emotion and through which our personal psychological landscapes are built. These in turn determine how well or badly a place works – even economically, let alone socially or culturally – and how it feels to its inhabitants and to visitors. Technical disciplines – engineering, physical planning, architecture, surveying and property development – are important, but they are a smaller part of the urban story than their practitioners would like to think.

Marrakech: Islamic architecture tends to protect residents from view.

Rome: Piazza Navona - a sensual delight.

The world of urban strategy though is still too affected and dominated by the urban engineering approach to city making, with its focus on hardware and the physical, rather than creative city making which uses the combined insights of say cultural geography, economists, historians, writers, environmental psychologists, social activists, specialists in participation, anthropologists, ecologists, as well as physical planners, designers and architects. When these other disciplines are brought in it is usually as an add-on, rather than as equal partners.

To handle the complexity of the urban experience we need deep and discriminating minds.

MULTIPLE SENSES & INTELLIGENCES

The senses build the knowledge upon which our worlds are built. The sensory landscapes we focus on are the five senses first classified by Aristotle: Sound, smell, sight, touch and taste. Yet this list is not exhaustive. For example, perceptions of pain and of balance have been identified as distinct from these five.

Just as there are multiple senses there are several kinds of intelligence as Howard Gardner reminds us in his book, *Frames of Mind: The Theory of Multiple Intelligences*. For too long, he suggests, we have given greater credibility to the thinking intelligences concerned with numbers, logic and abstractions or words and writing. Sensory intelligences, on the other hand, have been given secondary status, such as the visual-spatial concerned with vision and spatial judgement; the body-kinaesthetic, concerned with muscular co-ordination and doing; and the auditory-musical, concerned with hearing and listening. Although we admire painters, singers and dancers, their insights are rarely incorporated into how the economic or social worlds could operate.

The kinds of imagination and thinking the arts engender by their focus on senses and sense-making rarely, if ever, carries into city-making. Increasingly, artists are members of planning teams, but still more as an exception rather than the rule. Usually, too, they are restricted to the visual, as in public art projects, where all too often artists are brought in as decorative embellishment and as an afterthought rather than as part of the initial conceptualisation of possibilities. Artists play larger roles in urban events, but little as healers of the soundscape or developers of colour strategies.

Yet understanding of the sensory should not just be the domain of artists. All those concerned with cities should be sensitised and learn to incorporate this knowledge into their work and especially the 'hard' disciplines such as engineering, accounting and finance or the legal profession, since their work so determines how cities unfold.

esloe, WA:
pture by the
exhibition.

THE POWER OF METAPHORS

A helpful device to bring our urban world closer to ourselves and our senses is to think of any city as if it were a person, in order to get rid the tendency to see it as a lifeless being. What kind of person is Aberdeen or Ghent, two places I have recently worked with?

Aberdeen with its 'granite city' image is enterprising with a problem solving intelligence as well as tenacity, common sense and resilience, yet little flight of fancy and not too flowery, an apparent lack of emotion or excitement, a certain understated quality, some say dourness, it is more modest than showing off. They contrast this to Glasgow, which they feel is more showy. Ghent is self-critical, independently minded, it does not care what people think of it, it wants to go its own way, it dares to be creative and is unafraid of building well beyond its industrial past. Equally, it feels it need not shout, whereas nearby Antwerp they believe, seems to need to constantly promote itself.

The same exercise with metaphors can be expanded. If a particular street were a piece of clothing, what would it look like? How would it taste if it were a dish of food? What recipe would you give the city, is it a stew or a fry-up? What book does the city remind you of? What film and what atmosphere in it? The game can be played endlessly. Yet the intent is serious, namely to reach a richer feel for the city and its prospects and potential.

Metaphors help us jump out of our pre-existing thought patterns and cut through clutter. At times they allow us to see things afresh. They help create new associations as well as reflection and learning. They can expand and enrich our experience of life. This is why Venice is so symbolic. By thinking about the city and its potential futures and severe challenges it stands for a cluster of urban problems and how we might deal them. It is the utmost urban metaphor for our fragile world.

VENICE

Venice is a good place to use to discuss the sensory landscape and especially in June when the Biennale begins. It is easy to put Venice down, to say it is a tourist trap and that tourists are killing the vitality of the city, or that it is expensive and that the interesting Venetians have left and that it is a museum.

Much of this is true. But we must not forget why Venice is special. It was the New York or Shanghai of the 13th century. It was innovative and strategically constructed in a lagoon; it was not created by a 'business as usual' approach. It remains a sensory delight.

The Biennale in 2009 had the theme 'Making Worlds' curated by Daniel Birnbaum. It spreads out from its traditional site in the Giardini with its pavilions, to the vast Arsenale complex and then throughout the city where some countries exhibit, or collateral events take place. Immersive like the city is, so were many of the artworks.

Venice reminds us of many global dilemmas and our overall fragility. With the city's beauty this becomes especially emblematic. The fact that it is a sinking city alerts us now to what is to come with global warming, it tells us how we can lose what we have. The tourist flood highlights how the lifeblood and identity of a city can be drowned and drained. Currently there are 8 million tourists, which is 100 times the resident population and it is scheduled to rise to 12 million as the population declines dramatically to 60,000. This makes it 200 tourists for each resident, the majority of which are day trippers who contribute little to Venice's prosperity. This is why the city needs to draw in finance from all sources. Venice thus highlights the battle between commercial values and cultural values. You arrive at San Marco and are greeted by some of the largest ads in Europe. Swatch next to the famous winged lion as you enter by water. Then a few steps further at the Bridge of Sighs, part of the Doge's Palace, is a gigantic Sisley advert. You are not sure how to interpret this. Is it part of the Biennale?

nice: This is
Bridge of
hs if you can
ognize it.

Is the sponsorship to refurbish these sites worth their defacement? Are they beautiful? Are they ugly?

In the Biennale itself the worlds created often merge the media. The two prize winners in 2009 were Tobias Rehberger, who won the Golden Lion, for reworking the Palazzo delle Esposizioni's bland cafeteria into a wild world of zigzags in neon yellow, orange and black-and-white stripes, and Natalie Djurberg, who won the Young Artist's Award, with her 'Garden of Eden'. The Arsenale begins with the spiritual beauty of Brazilian artist Lygia Pape's sculptural installation, in which gold metallic treads illuminated by floor lamps are stretched from ceiling to floor to create the effect of a monumental sunbeam piercing the darkness. Then, throughout the immense spaces, world after world follows, such as Pascale Merthine Tayou's recreation of a full-scale African village.

Venice: Tobias Rehberger's award winning café.

Venice: The broader
context of the
Bridge of Sighs.

With the Biennale and city in combination you ask, is this a Gesamtkunstwerk? This forces you to ask fundamental questions, such as, can the artistic sensibility ground us back to what really matters: Creating places of conviviality that are sustaining emotionally, environmentally, economically and culturally.

Venice viscerally reminds us of our global dilemmas and fragility.

THE CITY AS A COMMUNICATIONS DEVICE

The lived urban experience comes from a circular sensory cycle. This imparts information. Its inert structures, the mechanical activities and how people behave and come across send out messages, we receive them as individuals and collectively throw them back at the city.

In essence, the urban experience is a psychological experience. The physical, social and sensory environment deeply affects the health and wellbeing of individuals and communities. Beauty and ugliness impact on our behaviour and mental state; building configurations can engender feelings of safety and fear. People have thresholds of tolerance as to what they psychologically can bear in terms of stimuli.

Sensory resources and awareness are still seen as offbeat, without much credibility, and there are very few educational programmes that explore what they can do. In spite of differences about interpretation, there are broad agreements on the significance of the senses across time and culture.

Rich sensory places are vital, vibrant and usually viable. There are some criteria to measure this, these include: Critical mass is the starting point for vitality, it involves a density and wealth of activity, intense interaction, often formed into clusters and hubs. Then there is the need to foster creative potential by encouraging curiosity, openness to imagination, playfulness and humour, so a flexible, fluid form of being entrepreneurial can be expressed. This is turned into reality through innovative capacity, which harnesses this talent and turns inventions and ideas into innovations. Identity and distinctiveness anchors this activity, projecting the unique, the special, the characterful. This can be lifted to another dimension if it can be communicated iconically so that we grasp the totality of its intention in one, without the need for a step-by-step, linear, narrative explanation. This involves too, showing intent in the physical environment. So if being green is the aim, this needs to be expressed. Vitality rises with seamless

*irana: Hundreds
*f oddly painted
*uildings attempt to
*hange a once dull
rban landscape.

Berlin: The exterior of buildings tell a story.

connectivity as this creates a pressure-cooker of interactions, physically, socially, culturally, emotionally and even virtually. This linkage, synergy and accessibility is fostered by mixed use, which generates diversity that engenders richness, complexity and choice. Here a dynamic mainstream and counterculture can develop and the blending of the old and the new, whether in terms of activity or the physical fabric, is key. Yet this vibrancy is not a free for all; it needs a sense of security, stability, safety and trust or let us use the old word – civics. This is how urban citizenship develops in a public realm that is valued.

The result, and I have encapsulated this before like my own personal mantra, is we hope orchestrated by strategic agility:

• A place of anchorage

• A place of connection

• A place of possibility

• A place of learning

• A place of inspiration.

A dynamic mainstream and counterculture can blend well the old and new.

*ngdao: A traditional
ban quarter – how
ng will it remain?*

57

URBANISM & ENVIRONMENTAL PSYCHOLOGY

Urbanism is the overarching discipline, which helps us understand the dynamics, resources and potential of the city in a richer way. It comprehends the physical, the sensory and human dynamics in a city and how they interrelate: In other words, the hardware and software priorities.

Seen so, urbanism is the most important 'urban discipline' and urban literacy is a linked generic and overarching skill involving the ability and skill to 'read' and 'decode' the city in all its tangible and intangible dimensions, including the sensory. A full understanding of urbanism only occurs by looking at the city from different perspectives. By reconfiguring and tying together a number of disciplines, penetrative insights, perceptions and ways of interpreting, an understanding of urban life emerges. Whilst people take disciplines like physical planning, geography, economics and even culture seriously in urban development, appreciation of the power and potential of sensory analysis remains very weak.

Given the link between the sensory landscape of cities and emotions and thus their psychological impacts on individuals and groups it is astonishing that environmental psychology is not a core urban discipline. Its focus is the interplay between people and their surroundings and the degree to which it creates stress or feels restorative. The discipline has a rich history stretching back over 50 years and assesses issues such as: The harmful effects of ugliness, whether a building, cheap materials or bad urban design. Equally it explores the impacts of signage clutter, dirt, over-stimulation and information overload, as well as how too much noise causes people to shut off and become uncommunicative. It touches on the disorienting effects of some urban environments and its consequences for feeling safe. It looks at the influence of height, heaviness and clunkiness on the senses and feeling overwhelmed or overshadowed by such townscapes. It

ew York: Broadway
d 42nd St. have
en made walkable by
amatically reducing
e space for cars and
providing places for
ople to relax.

draws out the consequences of seas of endless asphalt, car domination, physical barriers or sprawl on the sense of place, place identity, community building, civic pride, care and engagement. It shows how ugliness, a relative term of course, increases crime and fear of crime, leads to stress, vandalism, untidiness, feelings of depression, isolation, loneliness, worthlessness, a lack of aspiration and a drained will. The consequence is a self-reinforcing negative cycle. A core question, therefore to any architect or urban designer might be: 'how does my structure or plan help build social capital or revive the senses?'

Toronto: Daniel Libeskind's addition to the Royal Ontario Museum. His deconstructivist shapes reflect his claims that human existence is broken.

Qingdao: When cars dominate urban design people seem like a nuisance.

Thus, it concerns itself too with mental geography and how we orient ourselves and navigate the city. In sum, it looks at the thresholds people can psychologically bear.

It analyzes the reverse, namely the restorative effects of the old fashioned words such as beauty, even though what beauty is, remains contested, context driven and culturally specific. It will always be subject to debate. Yet this urban discussion is precisely the aim of good city making and the evolving urban culture of a place.

The interplay between people and their surroundings can feel stressful or restorative.

With all new fields the question remains: is creating this environmental awareness the role of a specific department or a sensibility that should be embedded across the city. Initially at least both approaches should be pursued. I have, unfortunately, so far, never met a high-ranking urban official who excels in sensory understanding and is an expert in environmental psychology.

Berlin: Cou
messages fo
on the side o
famous, squa
Tacheles buil

CHARLES LANDRY

Speaking, projects & research:
Charles Landry gives tailor-made talks on a wide range of topics, including: 'The Art of Great City Making'; 'Getting your City onto the Global Radar Screen'; 'Beyond the Creative City'; 'Punching Above Your Weight As A Smaller City'; 'Innovative Approaches To Running Complex Cities'; 'Imaginative Examples Of Developing Urban Cultural Resources'; and 'Green Urbanism'. These talks are aimed at a diverse range of audiences from city leaders to urban activists and the business and development community. Charles has recently spoken at events in Sydney, Shanghai, Abu Dhabi, Amsterdam, Milan, Minneapolis and Toronto.

Charles Landry undertakes task-specific projects or research, and acts as a commentator writing articles. He undertakes residencies ranging from one week to three months where a specific challenge or opportunity is worked through and presented publicly at the conclusion. While other projects involve deeper, longer-term relationships with cities and organizations: where strategies are created and tracked, reports are written and meetings chaired and facilitated.

The Creative City Index:
Charles Landry and Jonathan Hyams have developed The Creative City Index. This strategic tool assesses and measures the imaginative pulse of cities. To date, a dozen cities have been assessed including Bilbao, Perth, Canberra, Freiburg, Ghent and Oulu.

Books:
Titles by Charles Landry can be purchased online and from bookshops. Comedia Shorts are also available as e-books. Discounts are offered on bulk orders over 10 books, for events, workshops and meetings. **www.charleslandry.com**.

Contact:
To book Charles Landry as a speaker, to discuss a project and to enquire about The Creative City Index go to: **www.charleslandry.com**

Titles by Charles Landry:
Comedia Shorts 01: *The Origins & Futures of the Creative City*.
ISBN: 978-1-908777-00-3

Comedia Shorts 02: *The Sensory Landscape of Cities*.
ISBN: 978-1-908777-01-0

Comedia Shorts 03: *The Creative City Index*.
ISBN: 978-1-9087770-02-7

The Creative City: A Toolkit for Urban Innovators, Second Edition. Earthscan
ISBN: 978-1-84407-598-0

The Art of City-Making. Earthscan
ISBN: 978-1-84407-245-3

The Intercultural City: Planning for Diversity Advantage. Earthscan
ISBN: 978-1844074365

'Charles shows how new ways of thinking can help revitalize cities facing the challenges of the modern world'.

Sir Peter Hall